C000139579

Assigning Inorganic NMR Spectra

a programmed approach to solving NMR problems in main group chemistry

Table of Contents

Thanks

I have been supported in learning and teaching this topic by years of generous acts of collegiality from my teachers, my colleagues, and my students. Thank you all, friends.

I am particularly keen to thank those who commented on an early version of this book: Simon Aldridge, Chris Armstrong, Abbie Day, Jose Goicoechea, and Katherine Haxton. All errors remaining are mine alone.

If, as a reader, you have any suggestions about edits I should make for the second edition, please contact me directly. This is quite an exposed model of publishing (self-publication through Amazon's print-on-demand platform), and I am keen to improve things iteratively while keeping textbook costs low for students.

MON, Oxford, October 2019

michaeloneill.org

How to use this book

Inspired by Warren's *Chemistry of the Carbonyl Group,* I have structured this book as a 'programmed approach'. This format challenges you to solve problems through each chapter to construct your own understanding of the topic, and gives you feedback at every step. You can take things at your own pace. You can get things wrong before you get them right. No-one will ever know.

It is not intended to teach you everything, but it will help you gain a working knowledge of the important points in a complex topic. You should see it as one part of your Inorganic NMR education, alongside formal instruction and other textbooks. Some guidance on what to read next is given at the end of the last chapter.

You need to know a few things to use this book properly.

The whole book is split into *frames*. These are short sections separated by a page-wide horizontal line like this:

These lines are important because they map out a *sequence* of learning. You should physically cover up (with a scrap of paper or the cover of some lesser book) the frames you haven't reached yet.

You won't learn if you don't cover up the frames ahead of you; make good choices.

The text will occasionally have words in **bold**. **Bold** text is an instruction; you should do what the text says or answer the question it asks. You must do this in writing rather than just in your head.

This will force you to create (actually, physically create) your own answers rather than just scrutinise mine. The process of failing is extremely valuable in learning; if you find out you were wrong, comparing the answers will help you see how you can improve your thinking.

Warren's advice still rings very true: "it is often only when you commit yourself to paper that you find out whether you really understand what you are doing".

I hope you find the programme enjoyable and helpful. It's the book I wish I'd had when I was learning this topic.

MON, Oxford, October 2019

michaeloneill.org

A note to instructors

Assigning NMR spectra is a skill.

Not a fact. Not a concept. A skill.

Teaching skills is much harder than teaching other things. Students have to construct a set of linked procedures which help them respond correctly to a complex scientific artefact.

Broadly, my contention is that teaching this topic should help students to *assign* NMR spectra. When I was a student, I essentially learned to describe the relevant principles of NMR spectroscopy when non-hydrocarbon atoms got involved. These are very different things.

With this in mind, this book has taken a few bold decisions; I present them here transparently so you can decide whether O'Neill will sit well on your reading list.

1. The programmed approach helps students work through problems at their own pace. The whole pedagogy is driven by questions rather than answers. The appendices are mostly didactic, and the interludes are sort of half-and-half, but the chapters are pure problem-based learning.
2. I have arbitrarily restricted the chemicals. Specifically, there are no transition metal complexes. I hope the clear focus on main group systems reduces the cognitive load upon students as they start to grapple with a very difficult topic, and helps them to review VSEPR in a coherent way.
3. I have simulated the spectra: they are all 'fakes'. I feel this allows the principles to shine through much more clearly, but recognise that this decision will horrify some of you.
4. I have adopted a quantum number approach rather than the vector model throughout the book, though I have used the vector model in Appendix III, which addresses the physical basis of splitting.
5. I have aggressively shaved content which – in my experience – confuses students when they are still getting to know Inorganic NMR. If you teach a broader syllabus,

please be aware that this book might not cover everything. I personally like Weller and Young for the detail, but Rankin Mitzel and Morrison is also brimming with content. Iggo's OUP primer is a rich source of examples which align more neatly with a typical inorganic undergraduate course.

6. This book is not intended as a set of model answers, but a set of worked problems; your students will hopefully understand the principles of assignment from reading this book, but will need to be given chances to see how you want an answer presented in assessments.

If you, too, are aiming at helping your students to learn the practical skill of assignment, this book may complement your teaching. It's not a comprehensive text, but it is a short, focused introduction into the core aspects.

I note that it might prove a useful resource in developing flipped teaching, which is how I plan to use it myself.

MON, Oxford, October 2019

michaeloneill.org

Chapter 1: Transitions and Tree Diagrams

All spectroscopy is about *transitions*. In UV-Vis spectroscopy, this is normally the transition of an electron between orbitals. In IR spectroscopy, it is the transition between vibrational states in a molecule. In microwave spectroscopy, it is the transition between rotational states.

Nuclear magnetic resonance spectroscopy probes the transitions between atomic spin states.

All atoms have a nuclear spin quantum number, I (although this number can be zero). In a magnetic field, atoms have quantum spin states from $+I$ down to $-I$ in integer steps (so any atom has $2nI + 1$ states in total, where n is an integer). These spin states are described by the quantum number m.

Why do nuclei with $I = 0$ give not NMR spectra?

There is only one spin state ($m = 0$), so there is no *transition* between states to give an NMR signal.

The spin of the ^{1}H atom is ½, so it can exist in the $+$½ and $-$½ states (m values of $+$½ and $-$½). The spin of ^{2}H is 1, so it can exist in the $+1$, 0, and -1 states ($m = +1, 0, -1$). The *transitions between* these states is probed by NMR spectroscopy.

Why does the HF molecule have a ^{1}H NMR spectrum?

The ^{1}H atom in the molecule has two spin states, $+$½ and $-$½; NMR spectroscopy probes the transition between them.

Unlike most spectroscopies, NMR measures the radiation *emitted* when excited molecules relax rather than the radiation *absorbed* when they are excited.

Draw a diagram which represents the absorption-emission cycle.

You might have drawn something like this.

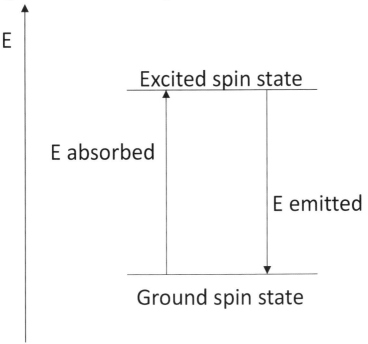

HF also has a ^{19}F NMR spectrum. **Why?**

The ^{19}F atom also has non-zero spin states, and so its transitions between states can be explored using NMR spectroscopy. ^{19}F, like ^{1}H, has $I = \frac{1}{2}$. **Modify our diagram from the last frame to show the explicit spin states**.

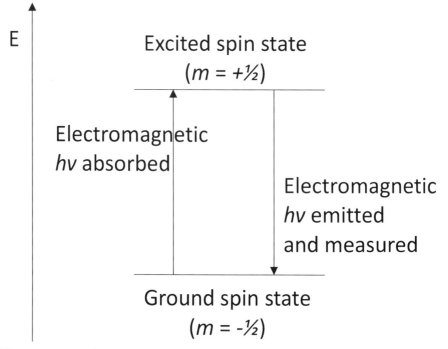

The energy emitted when the molecule relaxes is the most important factor in determining the chemical shift (δ). **How does this 'up-down' energy diagram relate to the 'right-left' NMR spectrum?**

The energy emitted is plotted right-to-left across the spectrum (δ can be considered a proxy for energy on any given spectrum; see Appendix II).

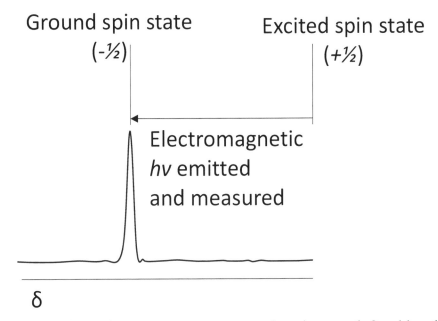

Note that the peak represents a *transition*: there is no peak for either the ground state or the excited state, only the gap between them.

How does this peak relate to the reference compound?

To properly calibrate machines, a readily-available reference compound is used to give a reliable fixed data point. For example, the ¹H reference compound is tetramethylsilane, which is *defined* as having a peak at $\delta = 0$. ¹H spin transitions with a greater energy gap than the TMS gap are referenced as positive shifts, while those with lower energy gaps are referenced as negative.

HF has a ¹H signal because the ¹H atom undergoes a transition from one spin state to another. The signal is a doublet. **Why?**

The ¹H atom *couples* to the ¹⁹F atom. The interaction of their local magnetic fields ('coupling') leads to a doublet *splitting pattern*. This can be linked to the transition model we have been developing if we note two things:

    1)   there is an evenly-distributed mix of spin states in solution at the start of the NMR experiment; and

2)   the magnetic transition happens for one atom at a time.

So the core ^{1}H transition can take place when either the ^{19}F is 'fixed' in either the $+\frac{1}{2}$ or $-\frac{1}{2}$ state for the whole excitation-relaxation process. The energy gaps associated with each ^{19}F context are slightly different, so the signal 'splits': half of the molecules report the ^{1}H transition when ^{19}F is in the $+\frac{1}{2}$ state. The other half report the transition when ^{19}F is in the $-\frac{1}{2}$ state. Appendix III has a deeper PhysChem discussion of this phenomenon.

This *probabilistic* model of splitting allows us to make use of an important conceptual tool: the probability 'tree' diagram. In its most basic form, a tree diagram can be used to show the probability of a flipped coin landing heads or tails.

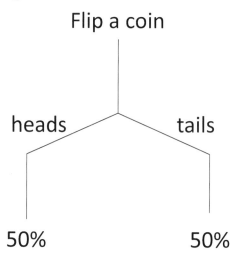

How can you use this model to predict the 1:1 doublet seen in the ^{1}H NMR spectrum of HF?

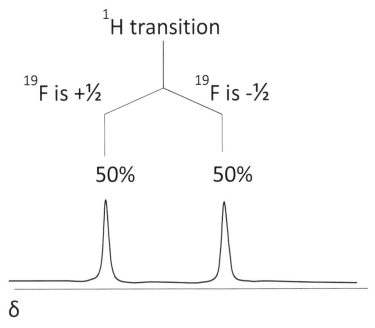

Note that this is not a single event, but a *statistical distribution* over all of the molecules in solution; we could compare this with flipping, say, 1000 coins:

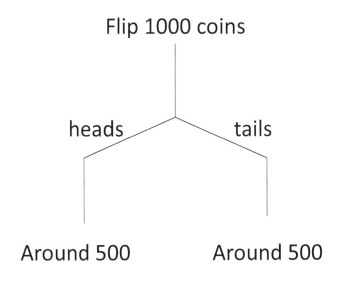

The chances of the ^{1}H transition happening when ^{19}F is in each state are 50:50, just like the coin toss. The relative intensities of the peaks represent the relative populations of molecules with ^{19}F in each of its (randomised) spin states.

The ^{19}F NMR spectrum of HF also shows a doublet. **Why?**

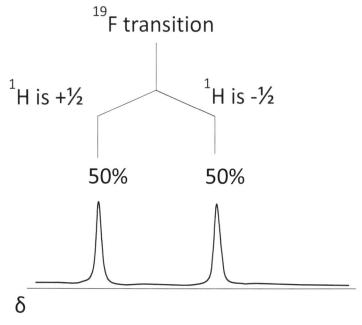

At the start of the experiment, there is an even population of spin states. When the magnetic pulse excites the ^{19}F atom, the ^{1}H population is split 50:50 between the $+\frac{1}{2}$ and $-\frac{1}{2}$ states. When the ^{19}F atom relaxes, the transitions in each case are different (so fall at different values of δ), but equally represented (so have identical integrals).

This way of linking probability 'tree diagrams' to observed coupling patterns is **the** core skill when assigning inorganic NMR spectra. Your main task is to become fluent in using tree diagrams to explain the patterns which your experiments or your examiners present you with.

The ^{31}P NMR spectrum of $PFCl_2$ is shown below. **Explain the splitting pattern.**

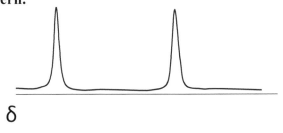

δ

Note: ^{31}P and ^{19}F both have $I = \frac{1}{2}$. You may assume that chlorine does not couple to any atoms.

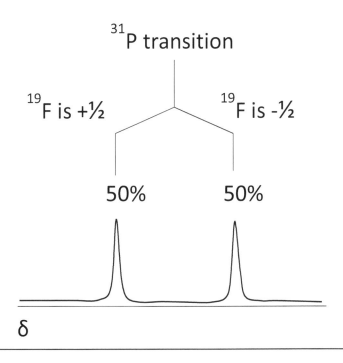

The ^{31}P signal is caused by the 'flipping' of the phosphorus atom's spin state, but the randomised spin of the ^{19}F atom influences the energy associated with this transition. Half the $PFCl_2$ molecules have 'spin up' fluorine ($m = -\frac{1}{2}$), while half have 'spin down' fluorine ($m = +\frac{1}{2}$).

The ^{31}P NMR spectrum of PF$_2$Cl is shown below. **Explain the splitting pattern.**

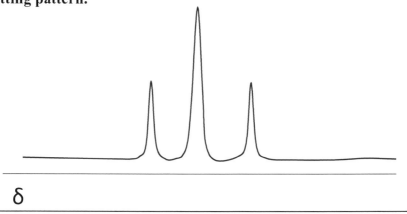

δ

If you are struggling, try thinking about the way that you might calculate the chances of getting a given result when flipping two coins.

The ^{31}P transition energy is affected by the independent randomisation of the ^{19}F atomic spin states. This adds 'branches' to our probability tree diagram.

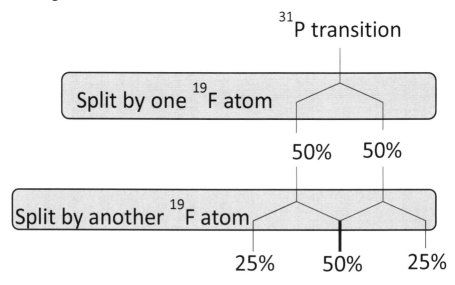

It is very important to note that the central peak of the triplet is made up of two overlapping peaks. This is the equivalent of flipping heads then tails compared to tails then heads: each fluorine atom can take m values of either $+ \frac{1}{2}$ or $-\frac{1}{2}$, so the possible fluorine spin configurations are

$$(+ \tfrac{1}{2}, + \tfrac{1}{2}), (+ \tfrac{1}{2}, - \tfrac{1}{2}), (- \tfrac{1}{2}, + \tfrac{1}{2}), \text{ and } (- \tfrac{1}{2}, - \tfrac{1}{2}).$$

You will note the similarity between this and the outcomes of tossing two coins:

$$(H,H), (H,T), (T,H), (T,T).$$

The middle two terms, as the F atoms exert symmetrical local magnetic fields on the phosphorus atom, result in degenerate transitions of the ^{31}P atom. This makes the central peak of the triplet twice as intense as the outer peaks, just like flipping a head and a tail is twice as likely a flipping two tails.

The ^{31}P NMR spectrum of PHF_2 is shown below. **Explain the splitting pattern.**

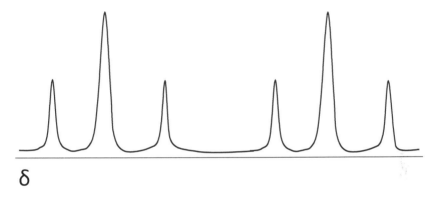

δ

Note: both ^{1}H and ^{19}F have $I = \frac{1}{2}$

If you are struggling, think about how a branched probability tree might look for three decision points (H, F, F). Which of these will show the same energy gap?

There are three independent splitting events. Each of these independently splits the signal given by the ^{31}P transition.

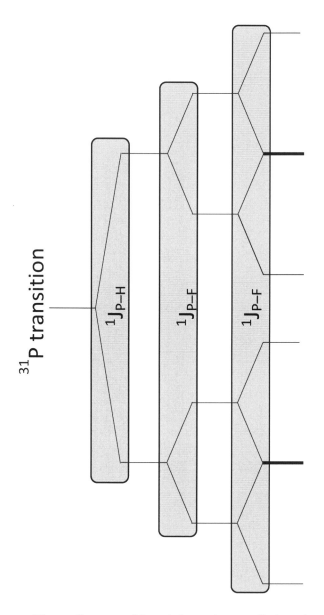

In my diagram, I have first considered the unique splitting due to the ^1H atom ($^1J_{P-H}$ means the 1-bond coupling between the P and H atoms). In any given molecule, it can be in either the $+ \frac{1}{2}$ or $- \frac{1}{2}$ state. Statistically, the PHF_2 sample will report ^{31}P transitions in each of these states in a 50:50 ratio.

The splittings due to each of the (symmetrical) F atoms are described next ($^1J_{P–F}$). The 1:2:1 triplet they create in the ^{31}P NMR spectrum depends upon the size of each splitting being exactly the same.

Note that the splitting of the ^1H atom is **not** the same as the splittings of the F atoms. This is because the local magnetic field exerted by an atom depends (among other things) upon its isotopic identity.

Note also that it doesn't matter which order you consider the splittings in; you might have addressed the fluorine coupling first and ended up with a 'triple doublet'. This is exactly the same as a double triplet. Physically, all the coupling is happening at once.

This concludes the chapter. By now you should have understood how to use this type of 'programmed approach' book, and grasped the core principles of tree diagrams. If you are still struggling with the diagrams, go back and work through relevant frames again – they are the core conceptual tool in this topic, and you need to develop mastery of them.

Interlude 1: Hz-ppm conversions

The gap between the peaks within a doublet splitting can be described in three different ways: the symbolic notation J_{a-b} (where a and b refer to the atoms involved in the splitting relationship); the difference in the chemical shift of the two peaks (in units of ppm); and the difference in precessional frequency (in units of Hz – see Appendix II).

Below are two NMR spectra of the same molecule. Each spectrum was run using a different magnetic operating frequency. **Why are coupling constants always reported in units of Hz?**

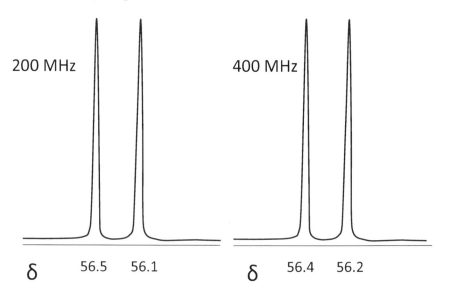

The coupling in ppm depends upon the strength of the magnet (see Appendix II). This means that the value of J in units of ppm is not a purely chemical property, but also depends upon the external physics of the instrument.

We solve this by scaling J by the external magnetic frequency (which varies, but is known for any given NMR experiment):

$$J \text{ in } Hz = J \text{ in } ppm \; \times \; Magnetic \; frequency \text{ in } MHz$$

You will notice that the mega (10^6) prefix of megahertz cancels the per million (10^{-6}) bit of parts per million, making the dimensions satisfyingly consistent.

Calculate J in units of Hz for the two spectra above.

In general

$$J \text{ in } Hz = J \text{ in } ppm \times Magnetic\ frequency\ in\ MHz$$

So for the 200 MHz spectrum

$$J \text{ in } Hz = 0.4\ ppm \times 200\ MHz = 80\ Hz$$

And at 400 MHz

$$J \text{ in } Hz = 0.2\ ppm \times 400\ MHz = 80\ Hz$$

These are the same number because the splitting in Hz is a property of the *molecule* rather than the magnet (though it is quite normal to see small disagreements in splitting values in the literature).

Calculate $^1J_{P-F}$ in units of Hz for the triplet seen in the ^{31}P NMR spectrum of PF_2Ph, shown below.

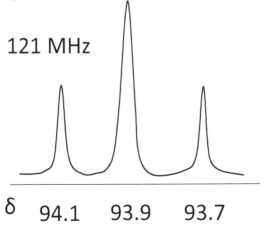

121 MHz

δ 94.1   93.9   93.7

A common mistake here is to mis-identify what J looks like on a complex splitting pattern. If you find yourself struggling, consider drawing out the tree diagram for this signal; a similar (triplet) diagram was presented in Chapter 1 if you want to look back.

The coupling constant spans the gap between two adjacent peaks (rather than the ones at each side. The gap is therefore 0.2 ppm (rather than 0.4 ppm):

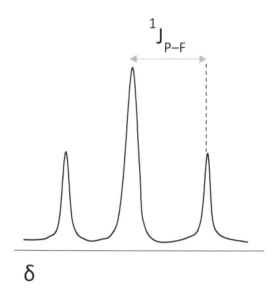

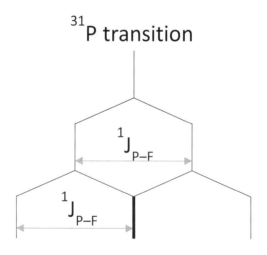

$$J \text{ in } Hz = 0.2 \text{ ppm} \times 121 \text{ MHz} = 24.2 \text{ Hz}$$

Hz-ppm conversion is an important skill, which is perhaps *fiddly* rather than difficult. In my experience, the most common problem is misremembering the formula (dividing the ppm splitting by the magnetic frequency). Second most-common is a simple numerical blunder (typing in the wrong number somehow). The third most common is mis-identifying what J looks like on the spectrum. This last one is a genuine conceptual error; the other two are probably less serious (especially if you show your working).

It is good to know some reasonable values for splitting constants. Organic H–H coupling is normally in the 1-100 Hz range. In my experience, numbers much smaller than this are an invitation to re-check which way up I put the formula.

In some ways, this whole problem is becoming rather inauthentic – most NMR software will do the calculation for you now. It remains a *really* nice exam problem, though. As an examiner, I have found it useful for adding a mark or two into a nearly-big-enough question.

Chapter 2: Chemical Environments

The low-resolution (i.e. with no splitting resolved) ^1H NMR spectrum of ethanol is shown below. **Explain the form of the spectrum.**

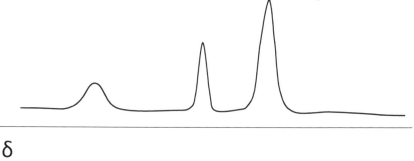

δ

Ethanol has three distinct types of hydrogen atom in it: OH, CH_2, and CH_3. The relative areas under each peak ('integrals' of each peak) help to suggest an assignment for the spectrum.

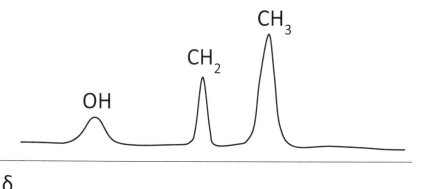

δ

Why does each hydrogen environment fall at a different chemical shift?

Each chemical environment has a different local arrangement of (spin-½) electrons, giving it a different local magnetic environment. The ^1H spin transition (which depends on the size of the local magnetic field) in each environment will therefore have a different energy. These different transition energies are observed as different chemical shifts.

The low resolution (i.e. with no splitting resolved) ^{19}F NMR spectrum of cold PF_5 is shown below. **Explain the form of the spectrum.**

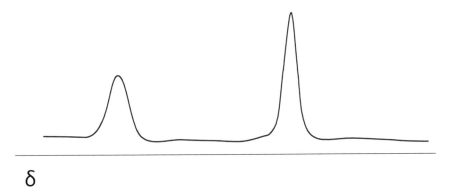

δ

Note: throughout this chapter, it will be useful to know that ^{19}F has a spin of ½.

If you are struggling, consider what shape PF$_5$ is. You may need to revise VSEPR (Appendix I).

The VSEPR-predicted geometry of PF$_5$ is trigonal bipyramidal. This gives two fluorine environments: two 'axial' fluorines and three 'equatorial' fluorines.

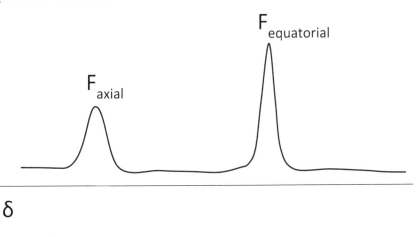

δ

The high-resolution (i.e. with splitting resolved) ^{19}F NMR spectrum of cold ClF$_3$ is shown below. **Explain the form of the spectrum using tree diagrams.** Try to do this on your own at first, but there are hints in the next few frames if you need help. You may assume that Cl does not couple to fluorine.

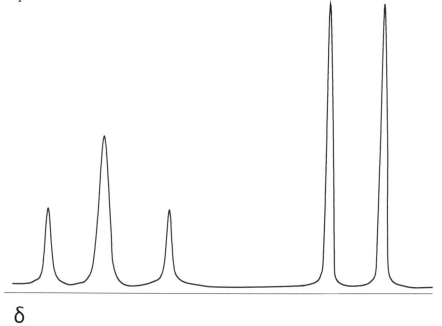

δ

First, you should use VSEPR to predict the structure of ClF$_3$.

Having determined that ClF$_3$ is 'T-shaped' (a trigonal bipyramid 'missing' two of the equatorial atoms), you should be able to identify two different fluorine environments.

For each of the two environments, you should now be able to construct a separate tree diagram based on what the environment 'sees'.

The doublet peak comes from two axial fluorine atoms coupling to one equatorial atom.

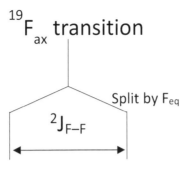

$^{19}F_{ax}$ transition

Split by F_{eq}

$^2J_{F-F}$

How do you explain the triplet signal?

The triplet is due to the spin state transition of one equatorial fluorine atom being split by two axial fluorine atoms.

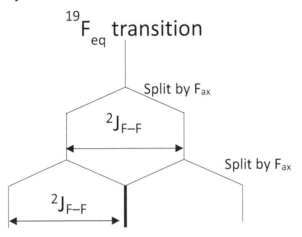

$^{19}F_{eq}$ transition

Split by F_{ax}

$^2J_{F-F}$

Split by F_{ax}

$^2J_{F-F}$

Note how this approach links the data to a proposed structure (the VSEPR T-shape) through the conceptual tool of the tree diagram.

The integrals of the peaks provide additional evidence to support this structure: the axial transition has a larger integral than the equatorial because there are twice as many ^{19}F atoms in axial positions.

Generally, chemical shift is less important in interpreting 'inorganic' NMR spectra than it is for organic ones; inorganic shifts are much harder to rationalise, and a greater emphasis is usually placed on using splitting patterns to assign spectra.

The high-resolution (i.e. with resolved splitting) ^{19}F NMR spectrum of cold SF$_4$ is shown below. **Explain the form of the spectrum using a tree diagram.** Try to do this on your own at first, but there are hints in the next few frames if you need help.

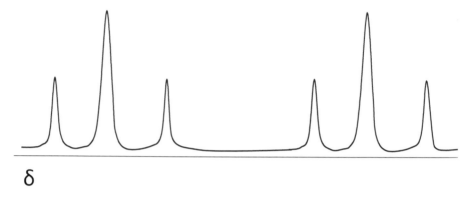

δ

First you should use VSEPR to predict the structure of SF$_4$.

Having determined that SF$_4$ is a 'sawhorse' structure (a trigonal bipyramid 'missing' one of the equatorial atoms), you should be able to identify how many fluorine environments there are.

For each of the two environments, you should now be able to construct a tree diagram based on what each environment 'sees'.

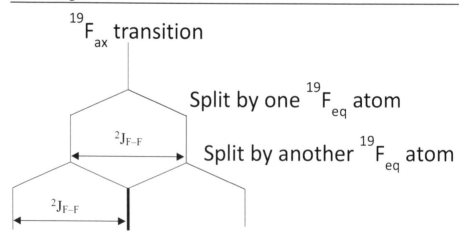

$^{19}F_{ax}$ transition

Split by one $^{19}F_{eq}$ atom

Split by another $^{19}F_{eq}$ atom

$^2J_{F-F}$

$^2J_{F-F}$

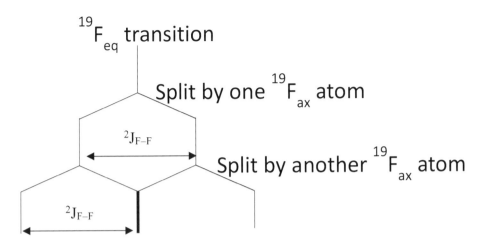

$^{19}F_{eq}$ transition

Split by one $^{19}F_{ax}$ atom

Split by another $^{19}F_{ax}$ atom

$^2J_{F-F}$

$^2J_{F-F}$

Note how the NMR spectrum is composed of *two distinct triplets*. Compare this with the double triplet you saw in the ^{31}P NMR spectrum of PHF_2 in the last chapter. Although these look similar, the conceptual interpretation is very different: SF_4 has two separate triplet signals in the ^{19}F NMR spectrum (which have the same integrals because each fluorine environment has the same number of fluorine atoms); PHF_2 has one signal (a double triplet) in its ^{31}P NMR spectrum.

The low temperature ^{129}Xe NMR spectrum of XeO_2F_2 is shown below. **Explain the form of the spectrum using a tree diagram.**

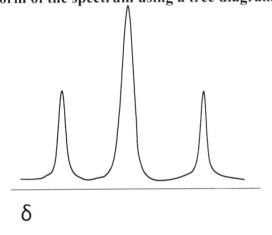

δ

The VSEPR structure of this compound is difficult (it is demonstrated in Appendix I). Remember that Xe=O double bonds are needed to help satisfy oxygen's octet, and that fluorine (the more electronegative group) will occupy the sites with the most crowding (because this structure holds the electron density away from the centre of the molecule).

The molecule is 'sawhorse' shaped again, with a linear F–Xe –F axis, and an O–Xe–O angle of around 120°. The two axial fluorine atoms split the ^{129}Xe transition into a 1:2:1 triplet.

^{129}Xe transition

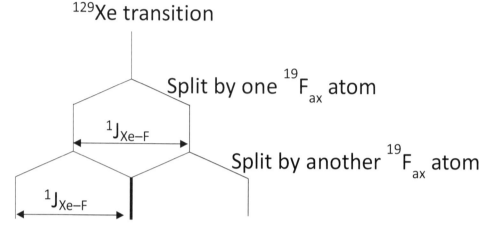

Split by one ^{19}F$_{ax}$ atom

Split by another ^{19}F$_{ax}$ atom

There is only one (triplet) signal, so this spectrum confirms that there is one fluorine environment in the molecule: both fluorine atoms are exerting symmetrical magnetic fields on the ^{129}Xe atom, each one splitting the signal by exactly the same amount.

The high-resolution ^{19}F NMR spectrum of cold PF$_5$ is shown below. **Explain the form of the spectrum.** Start (as ever) with a proposed VSEPR structure. Note that both ^{31}P and ^{19}F have $I = \frac{1}{2}$.

This is a difficult example, and you will find it challenging: take your time.

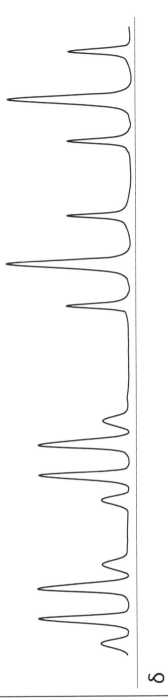

δ

As you determined earlier in this chapter, PF₅ is trigonal bipyramidal. There are two fluorine environments: axial (2F) and equatorial (3F).
What does each fluorine environment 'see'?

Each environment 'sees' two things: the central phosphorus atom, and the other fluorine environment. **Develop this description using a tree diagram.**

For the equatorial fluorine environment, the tree diagram shows a $^1J_{F–P}$ doublet coupling and also a triplet pattern due to the $^2J_{F–F}$ coupling to two axial ^{19}F atoms.

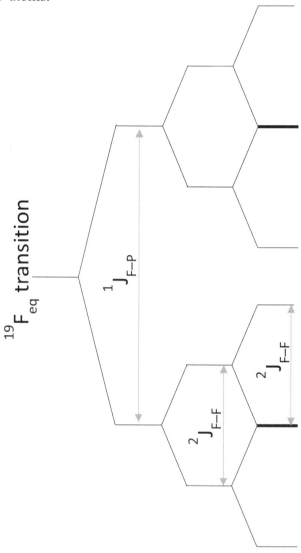

Construct the tree diagram describing splitting of the axial fluorine environments.

The phosphorus atom splits the signal into a doublet, and the three equatorial fluorine atoms carry out three symmetrical coupling events, producing a 1:3:3:1 quadruplet splitting pattern.

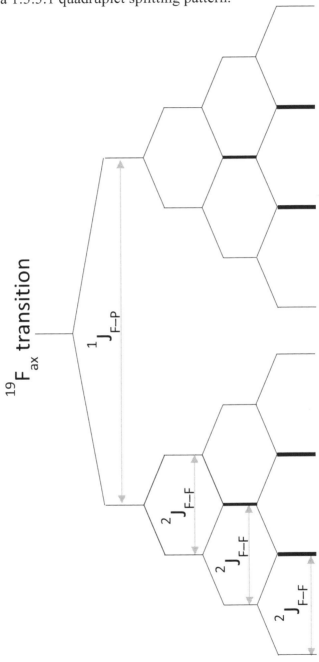

The 1:3:3:1 sequence is a classic example of the splitting described by Pascal's Triangle. This is constructed by working downwards, adding together the numbers directly above the space.

You may have encountered it when solving so called 'binomial' expansions of the form $(a + x)^n$, and it is common to hear Chemists talk about a 'binomial triplet' or 'binomial quartet', meaning that the peak areas follow Pascal's Triangle.

The triangle only works when the nuclei splitting the signal all have $I = \frac{1}{2}$ and are in the same environment. This specific situation arises a lot in organic molecules (e.g. the rapidly-rotating H atoms in a CH_3 group), but is much rarer in inorganic ones.

How does the NMR spectrum support the VSEPR structure?

Any NMR spectrum has a limited number of relevant features: chemical shift, splitting patterns, and integrals. The distinct chemical shifts of the 'double triplet' (dt) and 'double quartet' (dq) signals tells you that there are two fluorine environments. This is consistent with your axial/equatorial structure. The (approximately) 2:3 integrals support the number of fluorine atoms in each environment. Finally, the splitting is consistent with the number and identities of atoms 'seen' by each environment.

The key skill you need to master is analysing the structure of a molecule to identify the different symmetry environments of each type of nucleus. This is often easier to 'see' for organic molecules, where the differences between functional groups help make this process more intuitive (the ethanol example at the start of this chapter is a case in point). Small inorganic molecules, with their characteristic high symmetry, can take a little while to get used to.

Interlude 2: Decoupling

Coupling patterns are difficult things to learn about. This is because they have complicated relationships to extremely specific pieces of information about molecules.

This complexity is part of the value of NMR spectroscopy: it tells you things about the arrangement of atoms. But it can also tell you too much. Sometimes you don't want all that information.

How many carbon environments are there in propan-1,3-diol?

Propan-1,3-diol has two carbon environments. A classic 'carbon NMR' of this molecule is show below. **What is wrong with it?**

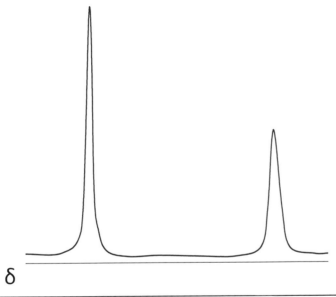

δ

The spectrum does not show any coupling to hydrogen atoms. This is completely standard practice, but involves a sophisticated process called decoupling. Specifically, the effects of coupling from ^1H isotopes are removed from the spectrum; the proton-decoupled carbon NMR spectrum is notated using curly brackets to identify the nucleus being decoupled: $^{13}C\{^1H\}$.

The ^{31}P NMR spectrum of PFH_2 is shown below. **Given the spin (I) and abundance (A) data provided, predict the form of the $^{31}P\{^{19}F\}$ NMR spectrum of PFH_2.**

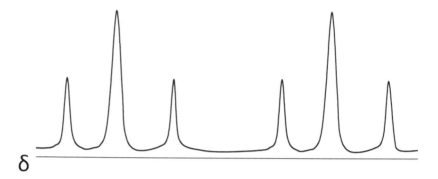

Note: ^{31}P has $I = \frac{1}{2}$ and $A = 100\%$. ^{19}F has $I = \frac{1}{2}$ and $A = 100\%$.

All coupling to fluorine atoms seen in the ^{31}P NMR spectrum vanishes in the $^{31}P\{^{19}F\}$ NMR spectrum, and a 1:2:1 triplet is observed due to the ^{1}H coupling.

The ^{31}P NMR spectrum of PF_2H is shown below. **Predict the form of the $^{31}P\{^{19}F\}$ NMR spectrum of PF_2H.**

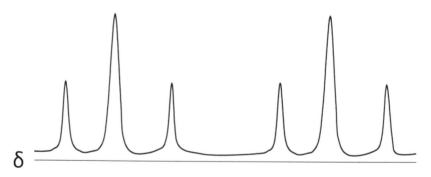

Note: ^{31}P has $I = \frac{1}{2}$ and $A = 100\%$. ^{19}F has $I = \frac{1}{2}$ and $A = 100\%$.

The $^{1}J_{P-F}$ coupling seen in the ^{31}P NMR spectrum vanishes in the $^{31}P\{^{19}F\}$ NMR spectrum, and a 1:1 doublet is observed due to $^{1}J_{P-H}$.

In most practical settings, this level of understanding is all you need to know: a decoupling experiment removes all coupling effects exerted by the atom targeted by the decoupling procedure. If you see $\{^1H\}$, ignore all the coupling from ^1H atoms. If you see $\{^{19}F\}$, ignore all the coupling from ^{19}F atoms.

Decoupling is physically carried out by a complicated series of radiofrequency pulses. If the details of these pulses are something your instructor thinks you should know about, then they will teach you about them. This book is intended to give you the skills needed to respond meaningfully to spectra (rather than a deep understanding of the underlying Physics).

Chapter 3: Quadrupolar Splitting

All the examples in this book so far have involved nuclei with a spin quantum number, I, of ½. ^{1}H has a spin of ½. ^{19}F has a spin of ½. ^{31}P has a spin of ½. However, other spin quantum numbers exist. The spitting patterns from coupling to nuclei with $I > \frac{1}{2}$ are particularly interesting, and are the main emphasis of this chapter. Nuclei with spins of 1 or greater are called *quadrupolar* nuclei.

CDCl$_3$, a common solvent, displays a characteristic 1:1:1 triplet in its ^{13}C NMR spectrum at around 77 ppm. This triplet is due to the carbon spin transition being affected by the spin states of the adjacent deuterium atom.

Deuterium (^{2}H) has $I = 1$. Any atom can adopt spin states (m values) from $+I$ to $-I$ in integer steps. **State the spin states which deuterium can take.**

$$+I = +1$$
$$+I - 1 = 0$$
$$+I - 2 = -1 = -I$$

So D can adopt three spin states: $m = +1$, 0, and -1. In general, an atom with spin I can adopt $2I + 1$ distinct spin states.

The 1:1:1 triplet seen in the ^{13}C NMR spectrum of CDCl$_3$ is shown below. **Explain the form of the splitting pattern by using a tree diagram.**

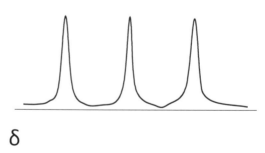

δ

The spin-1/2 tree diagrams we used in earlier chapters must be modified to account for the possible events. There is no longer a 50:50

probability of each outcome from a choice of two. Instead, there is a 33:33:33 probability of each outcome from a choice of three. **How can this be described using a tree diagram?**

The tree diagram should have three branches, showing the three spin states which the ^{2}H atom might be in when the ^{13}C atom relaxes from its excited state, emitting a measurable signal. The energy of each relaxation is slightly different, so the three transitions all fall at slightly different chemical shifts.

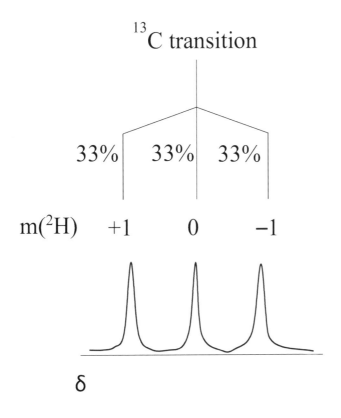

The ^{14}NH$_4^+$ ion displays a 1:1:1 triplet in its ^{1}H NMR spectrum. **Explain the form of the spectrum using a tree diagram.** Note that $I = 1$ for ^{14}N.

The four ^{1}H atoms are all in the same (tetrahedral) symmetry environment, so all have the same energy of transition (and same chemical shift). The energy released when a ^{1}H atom magnetically relaxes is slightly altered by the spin state ($m = +1, 0,$ or -1) of the ^{14}N nucleus. Each ^{14}N state is populated equally, so the relative probability of the ^1H transition happening in the three states is 1:1:1.

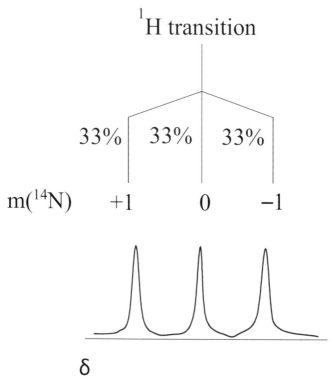

You will notice how similar this justification is to the ^{13}C NMR spectrum of $CDCl_3$. In both cases, the core transition is being split by one nucleus with I = 1 into a 1:1:1 triplet pattern.

---

^{11}B has a spin of 3/2. The ^1H NMR spectrum of $^{11}BH_4^-$ is a 1:1:1:1 quartet. **Explain the form of the splitting pattern by using a tree diagram.**

Hint: what values of m can a nucleus with $I = 3/2$ adopt?

This is another simple tetrahedral molecule. The one ^{1}H environment couples to the ^{11}B atom's $2I+1 = 4$ spin states.

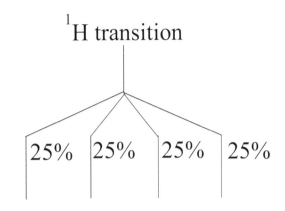

^{1}H transition

25% 25% 25% 25%

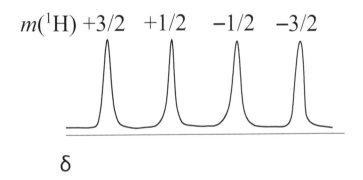

$m(^1H)$ +3/2 +1/2 −1/2 −3/2

δ

---

^{10}B has a spin of 3. **Why does the ^{1}H NMR spectrum of ^{10}BH$_4^-$ show a 1:1:1:1:1:1:1 septet?**

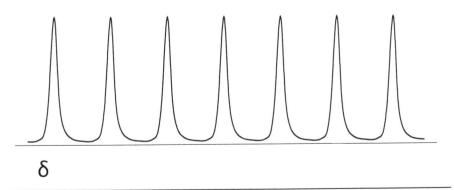

δ

Like the previous example, the H atoms couple to the central B atom. A nucleus with I = 3 can adopt the spin states (*m* values) +3, +2, +1, 0, –1, –2, and –3. Each of these states will give a slightly different energy for the ^{1}H nuclear spin relaxation (see Appendix III), so the spectrum shows a different peak for each of the ^{10}B spin states.

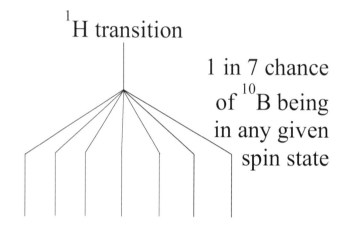

^{1}H transition

1 in 7 chance of ^{10}B being in any given spin state

The ^{11}B NMR spectrum of BH_4^- shows a 1:4:6:4:1 quintet. **Explain the form of the splitting pattern by using a tree diagram.**

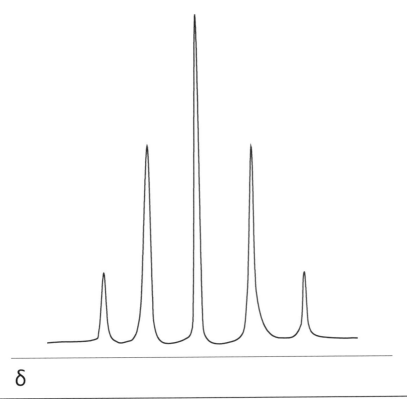

δ

The most common error students make in 'inorganic' NMR is mixing up which nucleus they are looking at. If you are struggling, think about which atom is undergoing the core transition and which atom is modifying this transition by splitting the signal.

The core ^{11}B transition is what is being measured, but the boron atom 'sees' four ^{1}H atoms. The electromagnetic wave emitted when the ^{11}B nucleus relaxes is split by the four ^{1}H atoms, each of which has $I = \frac{1}{2}$ (and therefore splits the signal into $2I+1 = 2$). Because these splittings are *exactly* the same, they overlap to give the 1:4:6:4:1 binomial pattern described by Pascal's Triangle:

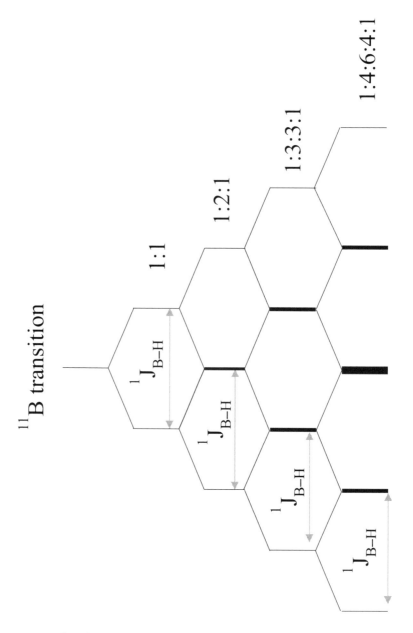

The areas under the peaks within a multiplet can only be justified by appeal to either the full tree splitting (the method emphasised by this book) or by appeal to Pascal's Triangle; remember that Pascal's Triangle *only* works when considering the splitting caused by spin-1/2 nuclei. By contrast, the tree diagram always works.

It can be hard to switch between thinking about spin-1/2 splitting and quadrupolar splitting.

---

^{73}Ge has a spin of 9/2. The splitting pattern in in the ^{1}H NMR spectrum of ^{73}GeH$_4$ is a 1:1:1:1:1:1:1:1:1:1 decet (10 peaks). **Explain the form of this splitting pattern.**

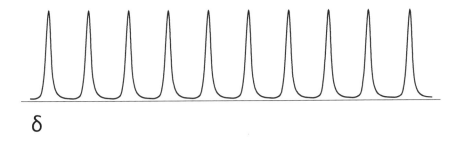

δ

Again, your tree diagram must be modified to embrace the possible spin states of the Ge atom through the hydrogen's spin relaxation. There are 2I+1=10 possible Ge spin states, reflecting values of m from +9/2 to –9/2 in integer steps.

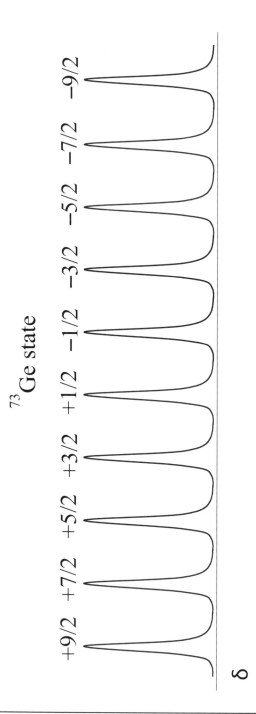

^{73}Ge state

+9/2 +7/2 +5/2 +3/2 +1/2 −1/2 −3/2 −5/2 −7/2 −9/2

δ

How should you report the chemical shift value of the ^{1}H NMR signal of this molecule. Why?

The reported shift should be the centre of any splitting pattern. Pragmatically, you might average the chemical shifts of two appropriate peaks to get the value. This is because the core hydrogen transition would fall at this shift in the absence of splitting (see Appendix III).

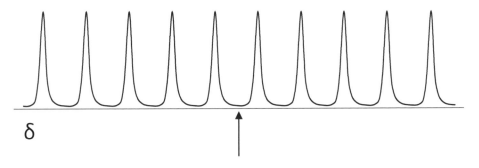

δ

Interlude 3: Intramolecular Fluxionality.

Molecules move. In solution, they translate, rotate, vibrate. Apart from translation, these motions have intramolecular consequences.

This is really important in NMR spectroscopy because it is a slow technique. Processes which involve motion much quicker than the timescale of a technique often result in complex artefacts which require delicate analysis.

You may have seen blurring in a time-lapse photograph. Interpreting the blur as (say) the passage of a moving car is quite intuitive to most people.

Interpreting the analogous 'blurred' features in an NMR spectrum is much less intuitive, but it's the same kind of problem: when it takes a technique (photography, NMR spectroscopy) a certain length of time to record something, things changing quicker than that time will be recorded in a time-averaged way (for the photograph: blurring) which reflects the nature of the change (the car moves, the molecule vibrates). This interlude is intended to describe what intramolecular motion 'looks like' on an NMR spectrum.

Trigonal bipyramidal molecules are often fluxional. Two different mechanisms of fluxionality have been proposed for this geometry: the 'turnstile' mechanism (left) and Berry Pseudorotation (right).

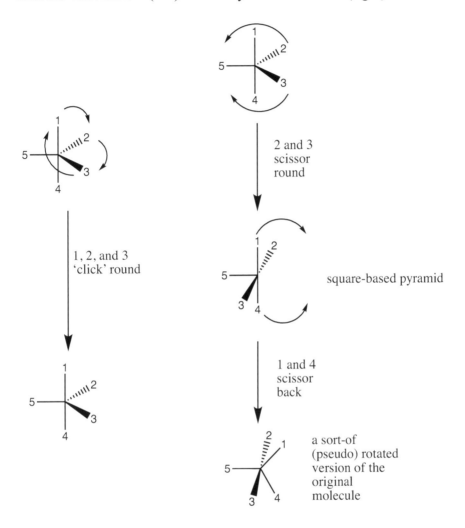

The precise details of the mechanism involved in fluxionality are not too important right now; the key thing is that *there is a way of switching atoms between axial and equatorial sites.*

At low temperatures, the SF$_4$ molecule has a ^{19}F NMR spectrum which derives from two distinct fluorine environments: there are axial and equatorial fluorines which are in different chemical environments.

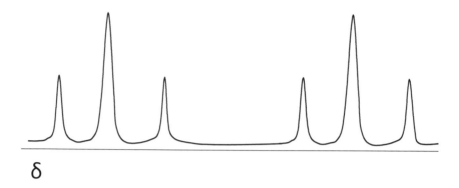

δ

The axial fluorine environment couples to the two equatorial fluorines (splitting into a triplet), and the equatorial fluorine environment couples to the two axial fluorines (splitting into another distinct triplet. They fall at different chemical shifts because they are surrounded by a different local environment of electronic charges (e.g. the axial fluorines have a right-angle orientation to the lone pair).

But when the system is heated up, the kinetic barrier to fluxionality is overcome by the thermal energy in the solution. This might be through the turnstile mechanism, or maybe through Berry pseudorotation – NMR spectroscopy can't tell us exactly how, but it can tell us that they interconvert.

Eventually, the rate of site switching will become fast enough that NMR spectroscopy (the very technique itself) cannot distinguish two distinct environments, but reports a blurred average environment. Being one environment, this will cause all fluorine atoms to fall at one chemical shift (between the shifts of the two distinct F_2 groups seen in the cold NMR spectrum). As there is no second environment, this F_4 group has nothing else to couple with; it presents as a singlet.

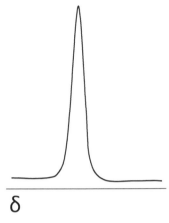

δ

This is something observed routinely in 'organic' NMR. The methyl group, CH_3, rotates so quickly in most situations that it is normal to think of it as containing one H_3 unit for NMR purposes; it might give a singlet in a ^1H NMR spectrum or cause 1:3:3:1 quartet splitting of an adjacent NMR-active nucleus' signal through a ^3J pathway.

Interlude 4 will describe how this qualitative discussion can be used to *quantify* kinetic parameters of the fluxional process by monitoring the precise details of when two signals merge; in general, this temperature of merging will be unique to the specific system.

Note that clues about fluxionality in exam questions usually involve temperature: a *cold* sample will normally have slower rates of interconversion, meaning you are likely to observe spectra associated with the rigid molecule. *Warm* or *hot* samples might have faster rates of interconversion, meaning that the NMR spectrum reflects the time-averaged environment.

VT-NMR leads to a (weirdly) common human problem. PhD students, hungry for knowledge about fluxionality in their molecules, often cool samples below the freezing point of their NMR solvent. If the solvent expands on freezing, this can shatter the delicate NMR tube. This is doubly heartbreaking because as well as losing their product, the student can't really blame anyone but themselves for not remembering the freezing point of water.

Chapter 4: Spin Dilution

The ^1H MNR spectrum of $[Na][^{11}BH_4]$ is shown below. **Assign the spectrum.**

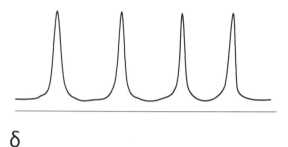

δ

Note: The spin of ^{11}B is 3/2.

The ^{11}B atom splits the signal given off by the magnetic transitions of the ^1H atoms. Any given ^1H transition might occur (with equal probability) in any of the $(2I + 1 = 4)$ spin states available to the ^{11}B atom ($m = +3/2, +1/2, -1/2, -3/2$). This example was discussed in Chapter 3 if you need to review this reasoning in detail.

The ^1H NMR spectrum of $[Na][^{10}BH_4]$ is shown below. **Assign the spectrum.**

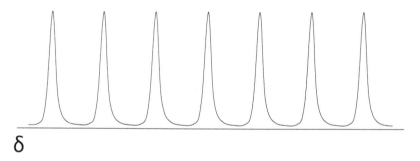

δ

Note: The spin of ^{10}B is 3.

Like the ^{11}B example, the different spin states available to the boron atom affect the energy of the ^1H transition. This boron atom just has more $(2I + 1 = 7)$ spin states ($m = +3, +2, +1, 0, -1, -2, -3$), spitting the ^1H signal more.

The ^{1}H NMR spectrum of reagent-grade $NaBH_4$ is shown below. **Assign the spectrum.**

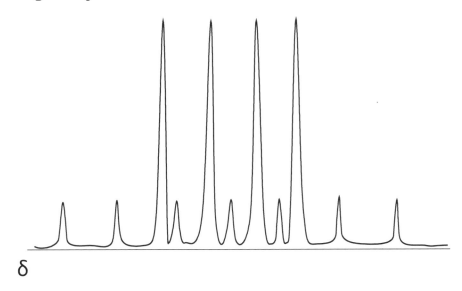

δ

Note: ^{10}B has $I = 3$ and $A = 20\%$; ^{11}B has $I = 3/2$ and $A = 80\%$.

Hint: use the spectra of the isotopically pure substances (which you've just assigned) to help you. **What mixture of molecules exists in solution?**

Hint: some (what proportion?) ^{1}H magnetic transitions occur in molecules containing a ^{10}B atom. Other ^{1}H transitions happen in molecules with an ^{11}B atom in the middle.

There are two different classes of molecule. 80% of molecules have ^{11}B in them, so 80% of the measured ^{1}H transitions reflect the 1:1:1:1 quartet splitting pattern due to ^{11}B. The 1:1:1:1:1:1:1 septet pattern due to ^{10}B is observed in proportion to the abundance of the isotope: 20% of the ^{1}H NMR spectrum reports hydrogen atoms in the $^{10}BH_4^-$ ion.

Why do both these patterns centre on the same chemical shift?

The ^{1}H NMR spectrum reports something about the ^{1}H atoms in the sample. All these atoms are in the same *chemical* environment (*i.e.* bound in the same way to the same element), so they all have the same value of *chemical* shift.

This is a specific case of a general phenomenon called *spin dilution*. The 'pure' spectra due to $[Na][^{10}BH_4]$ and $[Na][^{11}BH_4]$ are mixed ('diluting' each other like milk in coffee). This results in a weighted superposition of the spectra derived from pure components: the two 'pure' spectra are represented in proportion to their abundance in the sample, centred on the same shift.

In the reagent-grade $NaBH_4$ example, the ^1H NMR spectrum of $^{11}BH_4^-$ (which accounts for 80% of the molecules in solution) is split into a 1:1:1:1 quartet.

Each peak in the quartet therefore makes up 20% of the overall spectrum:

$$80\% \; of \; molecules \; \times \; \frac{1 \; peak}{4 \; peaks} = 20\%$$

What proportion of the spectrum does each of the 1:1:1:1:1:1:1 septet peaks represent?

$$20\% \; of \; molecules \; \times \; \frac{1 \; peak}{7 \; peaks} \approx 3\%$$

There is often a really serious issue in exam technique around spin dilution problems. Even students who understand the science often struggle to convey their grasp of the concepts. This is extremely normal, but also extremely frustrating (not only for you, but also for your examiners).

I *strongly* suggest explicitly calculating peak areas when you encounter Spin Dilution problems. Doing the calculation shows that you understand not only the abstract concepts of isotopic abundance and splitting patterns, but also how they interact in a complex context. Even if you get a number slightly wrong somewhere, you have demonstrated a clear grasp of the core concepts – we can give you marks for that.

The ^{1}H NMR spectrum of CH$_4$ is shown below. **Assign the spectrum.**

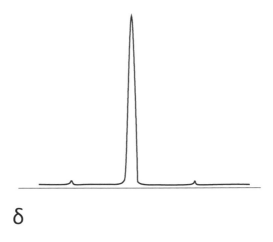

δ

Note: ^{13}C is 1% abundant and has a spin of 1/2.

The single proton environment mostly encounters ^{12}C, which has $I = 0$. Most of the spectrum represents this molecule through the 99% singlet.

But at the sides, a tiny doublet due to ^{13}C spitting can be observed (the distance between the small peaks is $^1J_{H-C}$); each of these peaks represents 0.5% of the overall spectrum:

$$1\% \ of \ molecules \ \times \ \frac{1 \ peak}{2 \ peaks} = \ 0.5\%$$

This 'satellite' splitting is a common feature of 'organic' spectra, but is often lost in the baseline noise. C–H coupling is generally exploited more systematically in 'two dimensional' techniques such as DEPT, which are outside the scope of this book.

People get three things wrong when assigning spin-diluted spectra.

1. They centre the patterns on different chemical shifts. This is wrong because the molecules are chemically identical, and so should have identical chemical parameters.
2. They do not weight the peak areas correctly. This part is mathematically quite fiddly, and requires some attention to detail (often in a stressful assessment situation).
3. They make a miscellaneous mistake. Usually this reflects either the stress of assessment and/or a lack of practice. Keeping track of everything requires real mastery of the topic, and mastery requires deliberate practice. Lots of it.

The ^{1}H NMR spectrum of GeH$_4$ is shown below. **Assign the spectrum. Note**: ^{73}Ge has $I = \frac{9}{2}$ and $A = 8\%$.

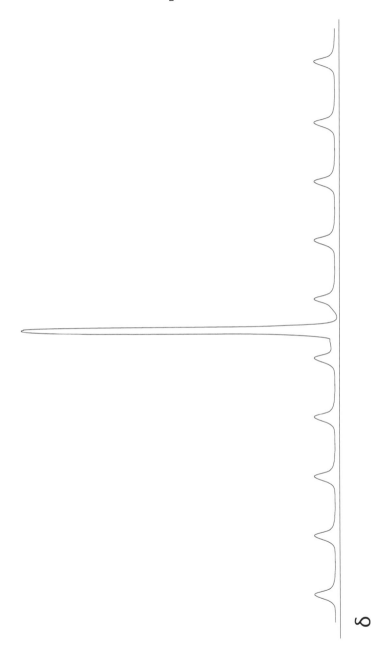

A very common question at this point is "what about the other 92%?"
Standard practice is to list (e.g. in a table) only those isotopes which
are spin-active (i.e. have $I > 0$). You can assume that other isotopes
(and there may be more than one) are spin-silent (i.e. have $I = 0$).

The big singlet is due to ^1H relaxations when the central germanium
isotope is spin-silent. This is true of 92% of the molecules in solution,
so takes up 92% of the spectrum.

The remaining 8% of the spectrum is split by the ^{73}Ge atom into a
decet ($2I + 1 = 10$); each peak represents 0.8% of the overall
spectrum:

$$8\% \; of \; molecules \; \times \; \frac{1 \; peak}{10 \; peaks} = 0.8\%$$

While there are no other signals for comparison, it is worth noting that
the overall signal integrates to 4H because there are four hydrogen
atoms in the single GeH_4 tetrahedral environment.

The ^{13}C NMR spectrum of the $Tl(CN)_2^+$ ion is shown below. **Assign the spectrum.**

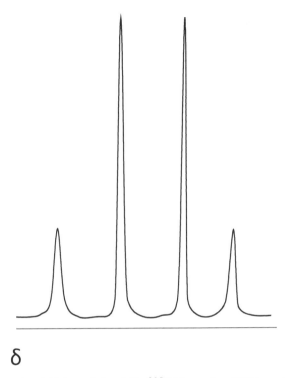

δ

Note: ^{203}Tl has $A = 30\%$ and $I = 1/2$. ^{205}Tl has $A = 70\%$ and $I = 1/2$.

The VSEPR structure for this ion is unusual: it ends up being a linear arrangement of the five atoms.

The one carbon environment gives a signal which reflects $^1J_{C-Tl}$ coupling. However, there are two different thallium isotopes, so some of the ^{13}C encounters ^{203}Tl and some of the ^{13}C is instead connected to ^{205}Tl. These are different isotopes, and so give different coupling constants. **What proportion of the signal is associated with one peak from each of these cases?**

For molecules containing ^{203}Tl:

$$30\% \; of \; molecules \; \times \; \frac{1 \; peak}{2 \; peaks} = \; 15\%$$

This (low) percentage indicates that these must be the outer peaks.

For molecules containing ^{205}Tl:

$$70\% \text{ of molecules} \times \frac{1\ peak}{2\ peaks} = 35\%$$

These must be the inner peaks, because the % is greater.

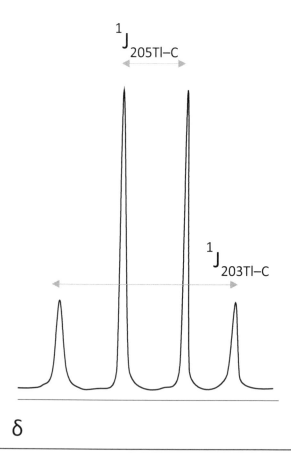

The ^{19}F NMR spectrum of the BrF_6^+ ion is shown below. **Assign the spectrum.**

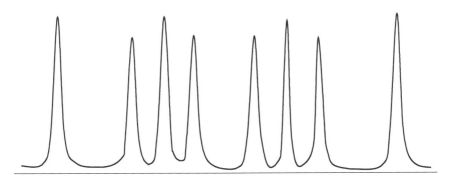

δ

Note: only $^1J_{Br-F}$ coupling is resolved. ^{79}Br has $A = 51\%$ and $I = 3/2$. ^{81}Br has $A = 49\%$ and $I = 3/2$.

Both Br atoms have a spin of 3/2, so will split the signal from the fluorine environment into four equal-area peaks because the ^{19}F relaxation can happen when Br is in any of its four spin states ($m = +\frac{3}{2}, +\frac{1}{2}, -\frac{1}{2}, -\frac{3}{2}$).

^{79}Br is slightly more abundant, so must be represented by the outer quartet; each peak in this system takes up 12.75% of the whole spectrum:

$$51\% \; of \; molecules \; \times \; \frac{1 \; peak}{4 \; peaks} = 12.75\%$$

^{81}Br peaks take up slightly less of the overall spectrum; they must be the inner quartet.

$$49\% \; of \; molecules \; \times \; \frac{1 \; peak}{4 \; peaks} = 12.25\%$$

This concludes the chapter. Spin dilution is very weird at first, but (eventually) slots in really nicely to some of the Chemistry you have studied in years gone by: the abundances of isotopes and the features that distinguish chemical phenomena from 'merely' physical ones.

This understanding is exemplified by the weighted superposition patterns seen in spin-diluted spectra, and can be communicated by explicitly calculating peak areas. Do this calculation in your exam answers: your marks will improve.

The last chapter is all about solving advanced problems using the skills you have developed already. These will be hard, but I hope you take a moment to reflect on how far you have already come.

Interlude 4: Variable Temperature NMR (VTNMR)

The last interlude discussed how chemical processes such as intramolecular fluxionality can be 'frozen out' by lowering the temperature of certain systems.

This allows a Chemist to investigate the fluxional process in some quite sophisticated ways. This Interlude will describe how kinetic parameters for a fluxional process can be derived from a series of careful NMR experiments.

At low temperatures, the chemical shifts of the two environments are distinct, and can be clearly discerned as separate signals, δ_A and δ_B. The *difference between* these shifts, $\Delta\delta$, is the first important piece of information:

$$\Delta\delta = |\delta_A - \delta_B|$$

The chemical shift difference, $\Delta\delta$ (in ppm), is not useful as a physical measurement for the purposes of kinetic studies, so it must be converted into a frequency, Δv in Hz, using the same relationship you used for coupling constants in Interlude 1:

$$\Delta v = \Delta\delta \times B_0$$

Or, in words:

$$frequency\ in\ Hz$$
$$= shift\ in\ ppm \times magnetic\ operating\ frequency$$

As the system is warmed up, fluxional processes bring the signals closer together (because they start to reflect a single time-averaged environment). The exact temperature that these signals merge together is the other important piece of information. For merging singlets, this is when two 'shoulders' merge into one broad peak with only one local maximum (boxed peak).

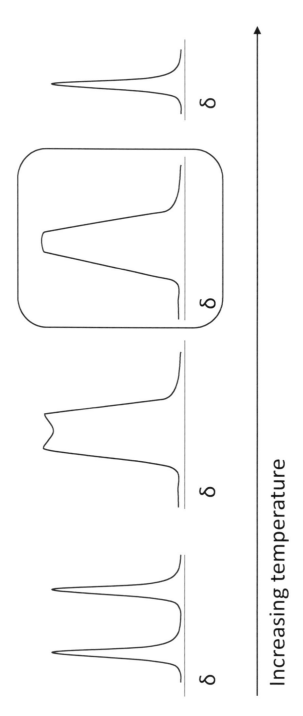

Interestingly, the spinner that you mount the NMR tube in is very important for this experiment. For wide temperature ranges, you should use a ceramic spinner because it does not expand or contract much with changing temperature. Normal (plastic) spinners can expand when cold, getting stuck in the machine and *really* annoying whoever has booked the instrument after you. Melting plastic spinners is less likely, because the supercooled NMR magnets can't really reach temperatures that high.

At the exact temperature when the two signals merge into one signal, the rate constant can be represented by the expression below. **Note**: this assumes first-order kinetics and that the merging signals:

(i)   have equal intensity; and
(ii) do not couple to each other.

$$k = \frac{\pi \Delta \nu}{\sqrt{2}}$$

This is a really important piece of data, because if you know the rate constant at a certain temperature, you can access the grand arsenal of classical kinetic theory.

Specifically, the Eyring equation can be used to tell you about the *activation energy* of the fluxional process:

$$k = \frac{\kappa k_B T}{h} e^{\frac{-\Delta G^{\ddagger}}{RT}}$$

The unfamiliar term here is κ, the 'transmission coefficient'. By my reading, this quantifies the non-energetic difficulty in crossing the barrier (e.g. orientations of collision/vibration). In the absence of information to the contrary, it is common practice to assume that this is equal to one.

The Eyring equation is then adapted by doing two things. First, the Gibbs energy of transition is split into the entropy and enthalpy of transition:

$$\Delta G^{\ddagger} = \Delta H^{\ddagger} - T \Delta S^{\ddagger}$$

Remembering that

$$e^{a-b} = e^a \times e^{-b}$$

This gives

$$k = \frac{\kappa k_B T}{h} e^{\frac{\Delta S^{\ddagger}}{R}} e^{\frac{-\Delta H^{\ddagger}}{RT}}$$

The second thing is to rearrange the whole formula in a *really* weird way. This is elegant, but also extremely contorted. The whole purpose is to get an equation of the form $y = mx + c$. This straight-line plot allows scientists to read numerical values for m and c straight off a graph (good). The problem is that the Eyring equation is impossible to express in this way (bad).

The solution is to mash the equation using a faintly horrifying series of logarithms and inversions.

First, divide through by T:

$$\frac{k}{T} = \frac{\kappa k_B}{h} e^{\frac{\Delta S^{\ddagger}}{R}} e^{\frac{-\Delta H^{\ddagger}}{RT}}$$

Now take the natural log of both sides:

$$ln\left(\frac{k}{T}\right) = ln\left(\frac{\kappa k_B}{h} e^{\frac{\Delta S^{\ddagger}}{R}} e^{\frac{-\Delta H^{\ddagger}}{RT}}\right)$$

And apply a rule about logs:

$$ln(ab) = ln(a) + ln\,(b)$$

To give

$$ln\left(\frac{k}{T}\right) = ln\left(\frac{\kappa k_B}{h}\right) + ln\left(e^{\frac{\Delta S^{\ddagger}}{R}}\right) + ln\left(e^{\frac{-\Delta H^{\ddagger}}{RT}}\right)$$

And a second rule about logs

$$\ln(e^a) = a \ln(e) = a$$

To give

$$\ln\left(\frac{k}{T}\right) = \ln\left(\frac{\kappa k_B}{h}\right) + \frac{\Delta S^{\ddagger}}{R} - \frac{\Delta H^{\ddagger}}{RT}$$

Which can be (linearly) plotted on a graph with a y-axis of $\ln\left(\frac{k}{T}\right)$ and an x-axis of $\frac{1}{T}$.

The line has a gradient of $\frac{\Delta H^{\ddagger}}{R}$ and a y-intercept of $\ln\left(\frac{\kappa k_B}{h}\right) + \frac{\Delta S^{\ddagger}}{R}$. This allows you to quantify the enthalpy and entropy of activation for a fluxional process.

A note to end on: this is an intimidating undergraduate task (especially in an exam), but in research Chemistry could well take a week of work. The slow pace of finding things out will allow you to properly understand what's happening if you ever have to do it 'for real'. Find out from your examiner (actually ask them) if this is an examinable topic in an exam setting. If it is, ask them if you have to remember any formulae or whether the final equation above will be provided.

What makes an NMR assignment hard to do?

You might want to think back on the times through this book when you have got things wrong. Was there a particular type of question you found troublesome?

Your answer to this question is more important than mine, but in my experience of teaching this topic there is one characteristic which unites the times students struggle: they all involve not just one feature of inorganic NMR (environments, quadrupolar splitting, and spin dilution), but at least two.

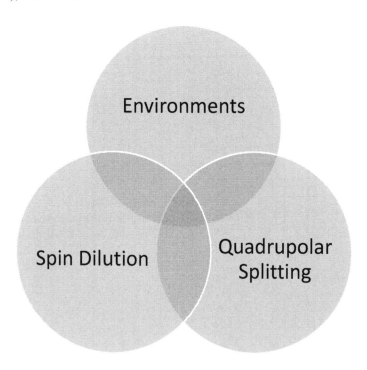

My conclusion from this observation is that learning how to assign such spectra is a *separate* step in your education; you need to practise assigning spectra which have more than one thing going on. This chapter aims to help you do that.

The ^{129}Xe and ^{31}P NMR spectra of [F–Xe–F–Xe–F]$^+$[PF$_6$]$^-$ are shown below. **Assign the spectra.**

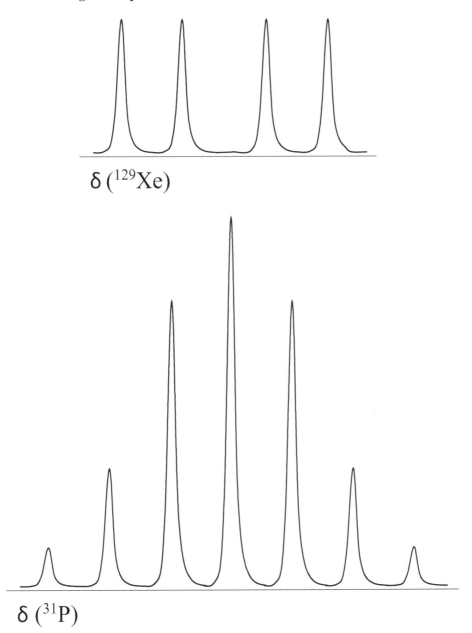

$\delta\,(^{129}\text{Xe})$

$\delta\,(^{31}\text{P})$

Note: You may assume that only ^1J coupling is resolved. ^{19}F has $I = \frac{1}{2}$ and $A = 100\%$. ^{31}P has $I = \frac{1}{2}$ and $A = 100\%$. ^{129}Xe has $I = \frac{1}{2}$ and $A = 26\%$.

The cation is an unusual V-shaped structure of C_{2v} symmetry. The ^{129}Xe NMR spectrum reflects the Xe atoms coupling to both the bridging ^{19}F atom and one terminal ^{19}F atom (as only one-bond coupling is observed). Each interaction results in a different doublet splitting, so the overall spectrum shows a double doublet (dd):

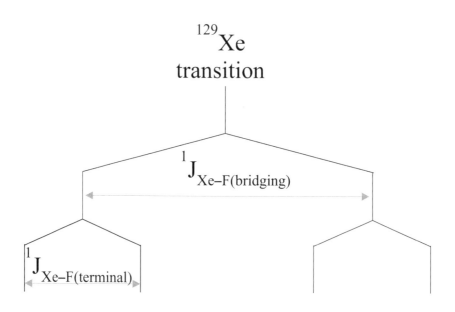

A common difficulty here is distinguishing exactly which nuclei are being considered; you have not been presented with the ^{19}F NMR spectra, and yet it is extremely normal to try and figure out what those would look like. This isn't even wrong, but it's not what you've been asked to do.

The ^{31}P NMR spectrum displays the binomial (Pascal's Triangle) splitting expected for coupling to spin-1/2 nuclei. Here, the phosphorus atom couples to six equivalent ^{19}F atoms, giving a 1:6:15:20:15:6:1 septet ($2nI + 1 = 2 \times 6 \times \frac{1}{2} + 1 = 7$). The tree diagram for this is quite big; using Pascal's triangle is much more convenient. Remember that Pascal's triangle *only* works when considering splitting to spin-1/2 nuclei.

Again, you have not been asked to assign the ^{19}F NMR spectrum.

The ^{129}Xe (top) and ^{19}F (bottom) NMR spectra of XeO_2F_2 are shown below. **Assign the spectra.** Note: ^{19}F has $I = \frac{1}{2}$ and $A = 100\%$; ^{129}Xe has $I = \frac{1}{2}$ and $A = 26.4\%$; ^{131}Xe has $I = \frac{3}{2}$ and $A = 21.2\%$

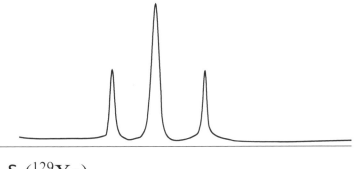

δ (^{129}Xe)

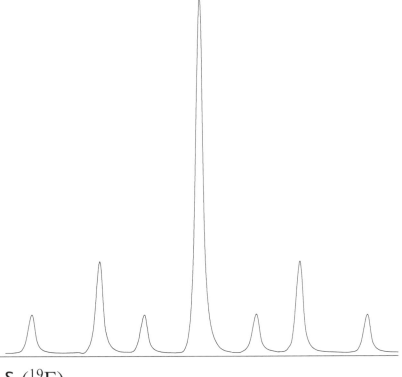

δ (^{19}F)

The molecule adopts the 'sawhorse' shape shown in Appendix I. Make sure you understand why the O atoms sit in the equatorial positions and the F atoms occupy the two axial sites of the trigonal bipyramid.

The ^{129}Xe relaxation is split by two ^{19}F atoms with exactly the same value of J, leading to a classic 1:2:1 binomial triplet.

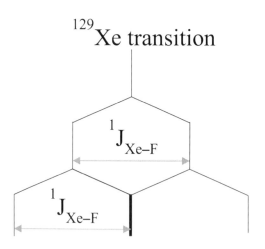

The ^{19}F NMR spectrum shows a couple of features, but all of them are bounded by the spin dilution effects which develop from the mix of Xe isotopes in solution.

^{129}Xe has a spin of ½, so will split the ^{19}F NMR signal in two; ^{19}F can relax when either m(^{129}Xe) = +1/2 or –1/2. This occurs in the 26.4% of cases where the molecule contains ^{129}Xe, giving each peak in the doublet an area of 13%:

$$26.4\% \; of \; molecules \; \times \; \frac{1 \; peak}{2 \; peaks} = 13\%$$

^{131}Xe has a spin of 3/2, so splits the ^{19}F signal into a 1:1:1:1 quartet ($2I + 1 = 4$). This is true of 21.2% of molecules in solution, giving each peak an integral just over 5% of the overall spectrum:

$$21.2\% \, of \, molecules \times \frac{1 \, peak}{4 \, peaks} = 5.3\%$$

^{other}Xe is presumably spin-silent, but accounts for the remaining 52.4% of the spectrum. In this situation, the ^{19}F atoms relax without being split by the central Xe atom. It is possible to formally calculate the integral associated with this event using the same equation if you want to; this signal should be the dominant feature of the whole pattern:

$$52.4\% \, of \, molecules \times \frac{1 \, peak}{1 \, peak} = 52.4\%$$

The observed ^{19}F NMR spectrum is a weighted superposition of each of these situations, because among the fluorine atoms in solution each coupling event happens in proportion to the abundance of the various xenon isotopes.

It can be hard to keep track of everything when you are being asked to look at different spectra of the same molecule. In 'real life', though, this is a very common way of using NMR spectroscopy. Demonstrating that several spectra support the proposed structure of a molecule is a convincing way to claim that you have made your molecule. Bolstering this with complementary data from other techniques (such as crystallography, UV, IR) can further support the rigorous argumentation which – for me – has always been the real intellectual thrill of synthetic chemistry.

The high-temperature ^{19}F NMR spectrum of BrF$_7$ is shown below. Assign the spectrum, assuming that only ^{1}J coupling is resolved.

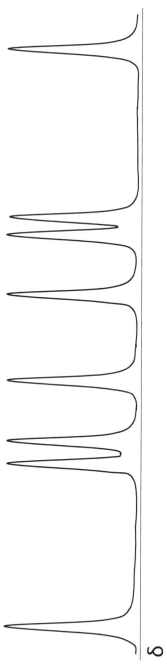

Note: ^{19}F has A = 100% abundant and has I = 1; ^{79}Br has A = 51% and I = 3/2. ^{81}Br has A = 49% and I = 3/2.

The ^{19}F signal is split into a 1:1:1:1 quartet regardless of which Br isotope is in the molecule (because both bromine isotopes have a spin of 3/2). With the abundances so similar, it is hard to use peak areas to identify the signals:

$$51\% \; of \; molecules \; \times \; \frac{1 \; peak}{4 \; peaks} = 12.75\%$$

$$49\% \; of \; molecules \; \times \; \frac{1 \; peak}{4 \; peaks} = 12.25\%$$

Without access to more advanced integration software, the best we can do here is to identify the two sets of signals. This can be a bit tricky, but rests on recognising that the splitting constants within a 1:1:1:1 quartet are the same no matter which two adjacent peaks you compare. **Try doing this using the symbols * and ^ to distinguish the overlapping 1:1:1:1 quartet peaks.**

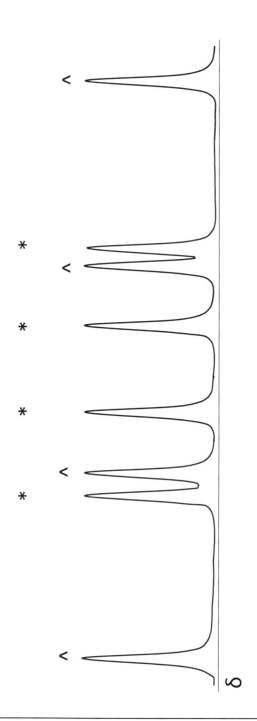

And that's it! You've finished the book.

This programme was intended to give you the basics, and you should now seriously consider using other texts to get further practice. I suggest Iggo's OUP primer *NMR Spectroscopy in Inorganic Chemistry* as the most focused source of material, but you should also consider Weller and Young's book *Characterisation Methods in Inorganic Chemistry*, which has some lovely discussion of the physics behind NMR and a clear discussion of how that develops into spectral patterns. Rankin Mitzel and Morrison's *Structural Methods in Inorganic Chemistry* is worth consulting if you do a masters project in Inorganic Chemistry, and worth buying if you do a PhD in it; it has a sharp focus on specific tricks which can be used in various special cases.

Inorganic NMR is really hard, until it 'clicks'. I hope that you have found this book helpful, and that you will contact me with your thoughts on it – good or bad. You need a lot of practice to 'get' this topic, but once you reach mastery it's a complete gift in assessments like exams: the questions generally don't take too long if you know what's happening, and it's a genuine disciplinary *skill* (rather than fact-heavy knowledge) which won't need much revision.

I strongly suggest that you make full use of your instructor's guidance:

- Listen to the things they emphasise in lectures;
- Think about the examples they use in their teaching;
- Ask how they design assessment questions for this topic (tell them the textbook you read said you should);
- Ask quick things before or after a lecture;
- If you have small-group teaching (and it's very common not to in later years of study as you become a genuinely independent learner), prepare questions before the session;
- Use resources like past paper questions to do assessment-aligned practice. If you can, do these timed to work out how you'll perform in the exam.

Good luck, wherever Inorganic NMR takes you. If you leave it behind forever (and most people do), I hope you see the conceptual difficulty of this topic as a really worthy part of a rigorous education.

MON, Oxford, October 2019

Appendix I: VSEPR Primer

Valence Shell Electron Pair Repulsion (VSEPR) is a model used to predict the shapes of molecules of p-block elements. The guiding principle is that outer-shell ('valence') electrons arrange their positions in space so that electron pairs are as far from each other as possible.

First, the model requires a Chemist to identify the central atom and count the number of electrons around it.

The central atom is normally the biggest one. This atom contributes a number of electrons equal to its 'old-fashioned' group number (e.g. C gives 4, N 5). You can think of this as the atom contributing all of its valence s and p electrons, but none of its d electrons (which are too low in energy).

Coordinating atoms share the appropriate number of electrons to achieve their octet. For example, F has 7 outer electrons, so shares one electron; O has 6 outer electrons, so shares two.

Add one to the electron count for each negative charge on the overall molecule; subtract one from the electron count for each positive charge.

Convert the number of electrons into VSEPR pairs. Normally, this will simply involve dividing by two. However, there are two important exceptions.

1. π-bonds do not 'count' in this scheme. This can be understood as a $(\sigma + \pi)$ double bond being only one region of electron density, so repelling the other electron pairs similarly to the σ-only single bond. E.g. E=O bonds should be counted as one VSEPR pair, not two.

2. Radicals 'count' as a pair of electrons. Again, this can be rationalised as a radical being one region of electron density.

The number of VSEPR pairs determines the arrangement of electron pairs around the central atom. These shapes are listed in the table below, which should probably be memorised.

VSEPR PAIRS	SHAPE
1	Linear
2	Linear
3	Trigonal planar
4	Tetrahedral
5	Trigonal bipyramidal
6	Octahedral
7	Pentagonal bipyramidal

Sometimes you will need to decide where to place an atom. For example, SF_4 is based on a trigonal bipyramid but could have its lone pair in either the axial or equatorial site.

Lone pairs of electrons repel a little bit more than E-X bonds. This is because only one nucleus competes for them, so they are held closer to the central atom.

E-X bonds with electronegative X atoms repel a bit less than expected. This is because the electronegative X atom pulls electron density away from the E nucleus.

Using VSEPR, predict the shape of XeO_2F_2.

This is a hard example. Take your time and refer back to the procedure if you need to. The process is taken step-by-step below if you need help.

The central Xe atom has 8 electrons (Xe is in 'old' group 8).

The coordinating F atoms each share 1 electron to achieve their octet. [2 in total]

The coordinating O atoms each share 2 electrons to achieve their octet. [4 in total] This will give a Xe=O double bond.

$$8 + 2 + 4 = 14 \text{ electrons} = 7 \text{ pairs}$$

But the Xe=O π-bonds don't 'count':
$$7 \text{ pairs} - 2(Xe=O \ \pi\text{-bonds}) = 5 \text{ pairs}$$

The shape for 5 pairs of electrons is a trigonal bipyramid. **How are the atoms arranged around the five potential sites?**

In order of repulsive power, the groups are ranked

$$\text{Lone pair} > O > F$$

Considering the trigonal bipyramid, there are clearly two different environments for the coordinating atoms: two 'axial' sites and three 'equatorial'. These sites have different energies.

Which one has the higher energy? Why?

An axial site has three 'nearest neighbour' groups at 90° to it. An equatorial site has only two 'nearest neighbour' groups at 90°. By simple electrostatic arguments (shorter distance gives greater repulsion), the axial sites are therefore of higher energy than the equatorial.

Place the less-repulsive groups in the sites where they can reduce the repulsion most effectively. This means that the F atoms go in the axial sites, and the O atoms go around the equator together with the lone pair.

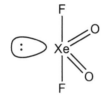

Appendix II: Energies of Magnetic Transitions.

Note: different instructors view the importance of this theoretical material differently. In my opinion, this material should be available in a digestible way to interested students but it is entirely possible to be an excellent synthetic chemist without a deep grasp of the quantum mechanics of nuclear transitions; I feel that the focus of learning NMR should be on practically assigning spectra.

Your instructor might take a different position, and I respect their judgement in the context of your specific educational setting. Make sure you work out how *your* examiner feels about this issue; the focus of assessments will always reflect their professional opinion about what's valuable to know.

The energy gap between two adjacent nuclear magnetic spin states in an atom is given by the expression

$$\Delta E = h v = h \frac{\gamma B_0}{2\pi} (1 - \sigma)$$

It is important to understand something about what some of these terms represent. The interpretation of v is perhaps the most complex part of the whole affair: it is the 'precessional' frequency of the nucleus.

Precession might be most familiar to you from astronomy. The earth spins on its axis once a day, but the axis is tilted. One precession occurs when the *axis* has gone through one cycle, tracing out a cone; this takes around 25,000 years for our planet. The 'wobble' of a spinning coin is a faster example of precession.

The gyromagnetic ratio, γ ('gamma'), is a parameter specific to an isotope. It can be thought of as a factor which quantifies the sensitivity of an isotope's precession frequency to an external magnetic field.

Tables of gyromagnetic ratios exist. The important thing to know for this topic is that they are *isotope-specific, not element specific*. For example, ^{10}B has a different gyromagnetic ratio to ^{11}B.

B_0 ('bee-nought') is a vector (shown by the **bold** typeface) describing the external magnetic field. In practical terms, this means the field exerted by the huge electromagnet in an NMR machine. These are typically dangerously powerful; 200 MHz magnets are fairly commonplace, with many departments having 400 or 500 MHz machines. You will normally see these fenced off with warning signs for people with pacemakers.

I ruined many watches during my PhD: I believe the NMR magnets magnetised the hands on the dial, slowing it down whenever they crossed each other. Importantly, this is not a metaphor.

The reason such powerful magnets are needed is that the other parameters are numerically very small once you've multiplied them by Planck's constant, h. To create a discernible energy gap, you need a big $\boldsymbol{B_0}$.

σ ('sigma') is the shielding constant, and can be either positive or negative. It quantifies how the local magnetic field scales the effect of the huge external field. This parameter is what makes NMR spectroscopy so useful to molecular scientists: local magnetic fields generated by the motion of electrons are highly dependent upon the spatial arrangement of atoms.

This spatial arrangement of atoms is often *exactly* what Chemists want to know. σ is the reason different chemical groups have different chemical shifts. More subtly, it is the reason for the 'splitting' phenomenon because the orientation of nearby *nuclei* also effects the local magnetic field.

Any quantised energy states can be treated statistically using the partition function, which describes the proportion of the population existing in the excited state at a given temperature:

$$\frac{n_{excited}}{n_{total}} = e^{\frac{-\Delta E}{K_B T}}$$

In the context of nuclear magnetic resonance, the expression for the energy gap between adjacent nuclear spin transitions can be substituted in for ΔE.

$$\frac{n_{excited}}{n_{total}} = e^{\frac{-h\gamma \boldsymbol{B_0}(1-\sigma)}{2\pi K_B T}}$$

Even with a very large $\boldsymbol{B_0}$, the energy gap between nuclear spin transitions is still very small because of Planck's constant. This leads to the approximately-even distribution of nuclei between the nearly-identical states.

$$\frac{n_{excited}}{n_{total}} \approx 1$$

This is the basis for the equal areas of the peaks in the doublet we saw in the ^{19}F NMR spectrum of HF in Chapter 1. The ^{19}F atom excited by a magnetic pulse has a 50:50 chance of being in a molecule with ^{1}H in the $+\frac{1}{2}$ state or the $-\frac{1}{2}$ state.

To repeat, different courses focus on the Physics of NMR to different extents. This book is aimed squarely at getting you to master *interpreting spectra*; the quantum mechanics takes more of a backseat. Be aware that your lecturers might focus the same topic to align more closely with PhysChem – I suggest bearing this in mind when you scrutinise past papers.

Appendix III: The Physical Basis of Splitting

The ^{13}C NMR signal for a ketone group falls at a different chemical shift to an alcohol group. **Why?**

The local magnetic environment around each ^{13}C atom is different. This affects the energy gap between the ^{13}C spin ground state and excited state through changing the value of σ discussed in Appendix II:

$$\Delta E = hv = h\frac{\gamma B_0}{2\pi}(1 - \sigma)$$

In this example, the particles exerting a local magnetic effect are the electrons arranged in bonds and lone pairs. **What other particles in atoms have magnetic properties?**

Nuclei (from the arrangement of protons and neutrons within the core of the atom) can also have magnetic properties. This is the basis of NMR spectroscopy.

The reason splitting happens is that the spin state, m, of nearby nuclei affects the energy levels of the nucleus under investigation.

For example the $^{31}P\{^{19}F\}$ NMR spectrum of $PFCl_2$ is a singlet. **What transition is responsible for this signal?**

Note: ^{31}P has $I = \frac{1}{2}$ and $A = 100\%$. ^{19}F has $I = \frac{1}{2}$ and $A = 100\%$.

Remember that the ^{19}F decoupling pulse – notated as $\{^{19}F\}$ – means that you can ignore any coupling to fluorine (see Interlude 2).

In this NMR experiment, the ^{31}P atom is excited from the lower to the higher spin state ($m = -\frac{1}{2}$ to $m = +\frac{1}{2}$) by the magnet. Relaxation from the higher state to the lower emits an electromagnetic wave; this relaxation-emission process is what generates the NMR signal.

The interesting thing about the effect of nearby nuclei is that the quantisation of energy levels means that there are a small number of specific events which can happen. At any instant in time, interacting nuclei can either align or anti-align, slightly affecting the energy of the state:

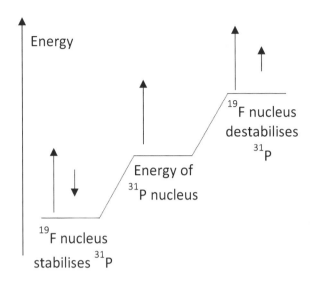

This is analogous to the way that bar magnets can either repel or attract each other when placed side-by-side.

The complication in NMR spectroscopy is that the signal is a transition between two states, rather than a single-state 'snapshot'. Quantum mechanical transitions obey certain ('selection') rules of symmetry, and the key one here involves the overall change in nuclear spin, ΔS:

$$\Delta S = \pm 1$$

In practice, this means that only one nucleus changes state in any single transition. This allows us to consider the ^{19}F atom as having a 'fixed' spin state (m) throughout the excitation-relaxation process of the ^{31}P atom.

The ^{31}P has two possible transitions rather than one. If the ^{31}P ground state is stabilised by the ^{19}F atom, the excited state is destabilised:

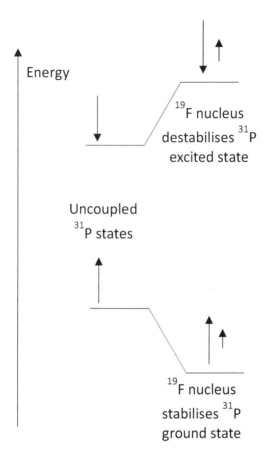

Energy

^{19}F nucleus
destabilises ^{31}P
excited state

Uncoupled
^{31}P states

^{19}F nucleus
stabilises ^{31}P
ground state

Draw the other transition, when the ^{19}F spin state destabilises the ^{31}P ground state.

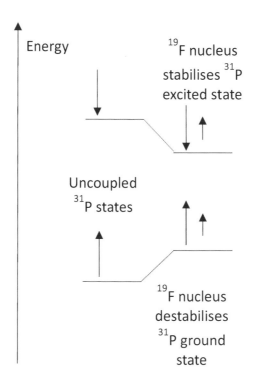

Energy

^{19}F nucleus stabilises ^{31}P excited state

Uncoupled ^{31}P states

^{19}F nucleus destabilises ^{31}P ground state

How do these two distinct transitions explain the doublet observed in the ^{31}P NMR spectrum of PFCl$_2$?

The energy of the ^{31}P transition is slightly modified by the adjacent ^{19}F spin state: there is now one slightly smaller gap and one slightly bigger one:

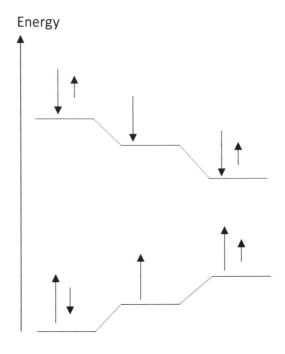

Energy

The two pathways are each followed by half of the molecules in solution, giving each peak in the doublet an equal area. The uncoupled transition is not observed, but would fall exactly between the two observed peaks at the same shift observed in the $^{31}P\{^{19}F\}$ NMR spectrum:

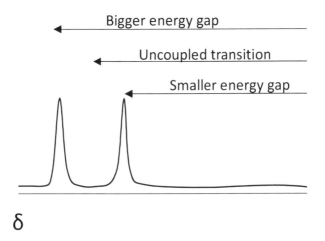

Bigger energy gap

Uncoupled transition

Smaller energy gap

δ

Reflections on Teaching Inorganic NMR

The first lecture I ever gave was a foundation year class on atomic structure. It was ok. My second, two hours later, was to third years on Inorganic NMR. It was *terrible*.

I made every mistake. I included too much content. I over-ran. I asked 'do you understand?' to the class (a question whose answer is apparently always 'yes'). I started from a mathematical consideration of the general principles of StatMech and slowly narrowed onto the specific application of NMR. I mumbled. I rushed. I encouraged students to hold their questions until the end, but then over-ran and had to leave so a colleague could set up for their lecture.

There are some specific things I wish I'd done from day one. This seemed like a good place to record a few of them.

Examine Authentically

This topic is being taught so that a student can one day muddle through a $^{31}P\{^1H\}$ NMR spectrum by looking back through their notes. They will never have to *predict* a spectrum, which was how I started off writing exam questions; I now see this like the difference between reading a foreign language and writing it. Students will never have to understand the detailed QM or StatMech when they do NMR 'for real'. Assessment should revolve around the core skill of assigning spectra, not ancillary detail. I got this very wrong at first, and my students didn't have the best shot at demonstrating their ability when I assessed that cohort. I regret it.

Line Up Teaching With Assessment

The nuts-and-bolts process of assigning spectra can be terrifyingly easy to avoid when lecturing didactically. Explaining, describing, enthusing - none of these lecturer behaviours quite match up with training the students to assign an NMR spectrum. I now try to spend contact time *actually, practically* assigning spectra rather than just *talking about* assigning spectra (or what spectra are in the abstract).

Trust That It'll Be OK

I over-prepared the first time I taught this course. This made my teaching sequence too rigid, and also too sensitive to my own experience of learning this topic (rather than the experiences of my students). I have found that my students learn best when I trust them to make use of me. This requires sessions with much looser structuring and structured student activities, but also the self-belief that I can cope with the questions they throw at me.

For this style of teaching to work, I have to actively make myself approachable; opening my lecture course with a discussion of the academic pressure of third year and a slide directing students to the University's welfare resources (including counselling) has really helped with this. I also include a few (*super* weird) cartoons from the A Softer World webcomic (asofterworld.com) in each session. These seem to provide a nice affective break from the Problem Solving.

I wish you every success in your learning and/or your teaching. There can't be many chemical skills harder to learn than assigning complex NMR spectra, and I hope you take a certain sort of pride in the scale of your challenge.

Good luck! And let me know if you find the book useful or can see ways to make it better.

MON, Oxford, October 2019

michaeloneill.org

Printed in Great
Britain
by Amazon